REPORT

OF

JOHN B. JERVIS,

CIVIL ENGINEER,

IN RELATION TO THE

RAILROAD BRIDGE

OVER THE

MISSISSIPPI RIVER,

AT ROCK ISLAND.

NEW YORK:
WM. C. BRYANT & CO. PRINTERS, 41 NASSAU ST., COR. LIBERTY.

1857.

REPORT

OF

JOHN B. JERVIS,

CIVIL ENGINEER,

IN RELATION TO THE

RAILROAD BRIDGE

OVER THE

MISSISSIPPI RIVER,

AT ROCK ISLAND.

NEW YORK:
WM. C. BRYANT & CO. PRINTERS, 41 NASSAU ST., COR. LIBERTY.

1857.

REPORT.

New York, 28th Nov., 1856.

To the Board of Directors of the Mississippi Bridge Company:

Gentlemen :

In compliance with a request from the President of your Board, (Henry Farnam, Esq.,) I have examined the bridge over the Mississippi, between Rock Island, in Illinois, and Davenport, in Iowa.

I was requested to examine the bridge, in reference to the structure and its adaptation to provide railroad connection between the States of Iowa and Illinois, and also its provision by a draw, to provide for the navigation of the river.

The great interests involved in the means of easy intercommunication, demand that the discussion of this subject should be entered upon in the most dispassionate manner.

The span that had been destroyed by fire in the early part of the season had been replaced at the time I examined the bridge, (26th September last,) and boats were daily passing through the draw.

The bridge appears well adapted to the passage of trains of railroad cars. Trains of twenty to thirty freight cars pass the bridge with the greatest facility. I observed, with great care, the effect of a train of thirty freight cars crossing it at the speed of about eight (8) miles per hour, with a heavy engine (25 tons), and could see no want of perfect stability. In its general appearance, the bridge is a substantial and beautiful structure. The severe ice of last winter passed the bridge without harm, and was a high test of its stability. Fire is the element that requires care to protect it from damage, and, no doubt, iron will eventually be substituted for the superstructure, both for safety and economy. This would be too costly at present; but the increase in the traffic will soon be such as to demand large expense, to protect it against the danger of interruption. Ordinarily there is not much danger from fire, and reasonable vigilance may preserve the structure for many years. The peculiar circumstances, by which one span was destroyed soon after the bridge was brought into use, may never occur again.

At this time, there are sixty-six (66) miles of railroad west of the Mississippi depending on this bridge for its eastern connection. The district thus accommodated has but a sparse settlement; and still the traffic that is brought to the bridge, and dependent on it for a rapid and cheap transit, has assumed an importance that could hardly have been anticipated, and would now feel very seriously any interruption in crossing. The present traffic, however large it may be regarded under the circumstances, is a trifle compared to what it must be, when the two lines of railroad that now concentrate (via Muscatine and Iowa city,) to this point, shall have crossed the State, and the growth of ten years in the settlement and improvement of the country shall have transpired. It may be regarded a question fully demonstrated, that the present and prospective traffic that will seek this avenue, fully warrants the maintenance of this bridge on a basis that will give the greatest regularity and certainty in its office. The practicability of the bridge, and its great economy and convenience to the public interests, can no longer be questioned.

In regard to the position of the draw in the bridge, I made all the examination I was able to make, without hydrographic survey of the river. Such a survey should be made, embracing about three-quarters of a mile, to a mile above and below the bridge. This would give a clear view of the channels, and show fully the course that steamboats must traverse in low water, and where they would be at liberty to move in high water. There is not much difficulty in judging by the eye, at low water, of the course boats could and would be likely to take, at a rise of six (6) to eight (8) feet above low water. Without a chart from survey as above mentioned, the low-water channel can only be traced by the currents on the surface, that are liable to mislead to some extent.

The opening of the east draw is one hundred and twenty-five feet (125), and the west draw is one hundred and twenty feet (120.) The east draw is the one most eligibly situated for the passage of vessels, and will no doubt be mostly used. The openings are ample for the class of vessels that usually navigate the river at this location.

It has been said that the piers of the bridge so obstruct the flow of the river, as to cause an increased current; thereby rendering it more difficult for steamboats to go up. When it is considered that the openings between piers (except at the draws) are two hundred and fifty feet (250), with piers of ten feet (10), it is obvious no material effect of this kind can be produced. A slight increase of current, no doubt, would be produced at high water, but not such as to produce any practical difficulty, as it is evident a steamer that could not pass up through the draw would be quite incapable of stemming the current in the rapids within a mile above. The only instance that I could learn, of any steamer having difficulty at the bridge, was the one that was burned. That steamer passed the draw and proceeded some two hundred feet above the head of the long draw-pier, and no indication of difficulty was manifested, until from some cause one of her wheels stopped, and the other being continued, the boat was necessarily turned round, and against the tendency of the current, and brought up against the bridge at the pier next east of the draw. In this position she lay, as I was informed, for some time, without indicating any special injury; the passengers and spectators called to the spot by the occasion, after delaying a while, separated and left the boat, no persons except the crew remaining. No indication of any danger to the vessel had yet appeared. Soon after this, the boat was discovered to be on fire, and communicated to the bridge, destroying one span. The fire is said to have originated from a stove in the cabin. The time that elapsed before fire appeared, the number of persons that were about the boat and in the cabin at the time and after the boat brought to against the pier, show conclusively that the fire did not arise from collision against the piers, or in any manner chargeable to the draw or bridge. All that can be argued from this circumstance against the bridge, is, that after the boat had failed in her machinery, she *might* have floated off in safety, had there been no bridge to strike; but this is not certain, for she might have floated against some other object, if there had been no bridge. I have been more particular in this case, to give the facts as near as I could ascertain them, for the reason, that the public have been led to believe the disaster to the boat was the result of some defect in the draw or bridge. The hardest point in the current is at the head of the draw-pier, and this the boat had passed about two hundred feet, with thus far no apparent difficulty. It is clear the boat changed her course (and this against the tendency of the current) by stopping one wheel, either by accident or design, and allowing the other wheel to work until the boat was rounded, and her head down stream. Had the other wheel been immediately or very shortly stopped, the boat might have floated back through the draw without harm from the pier. The fire undoubtedly originated from some other cause than the stoppage of the boat against the pier.

It has been complained that the line of the bridge piers are not directly in correspondence with the channel of the river. To a general observer on the shore of the river at Rock Island wharf, this appears to be the case, more especially at a high stage of water. It must be noticed, however, that the channel necessarily followed by vessels in low water often varies from the course of the banks or sides of the river. At a point about half a mile above the bridge, the low-water channel runs nearly at right angles with the general course of the stream for nearly half its breadth, and then turns almost at right angles, and apparently continues very nearly in direct line until it passes the bridge about a quarter of a mile, and then changes its course. Except as described to me by persons who had observed the course of boats in low water, I had no means of judging of the position of the channel, except as indicated by the superiority of the current. This led me to believe that the line of piers cut the channel at a slight angle, and I was led to consider whether this circumstance would offer any difficulty in navigating a boat through the draw. The width of opening at the draw is very liberal for the class of boats that navigate the river, even with the barges that they sometimes carry by their sides. The angle is very small between the course of the channel and the piers. The guide-pier is four hundred feet long, and, of course, effectually controls the current, bringing it in direct line with the pier as soon as it passes the head. It is, therefore, only necessary to navigate the boat so as to clear the head of the pier: after passing this point, the direction of the current tends to keep it in proper course to pass the draw. If a sharp turn of the boat was required, there might be difficulty; but the angle is so small, it is only necessary for the navigator to hold his rudder steady on one side, as would be indicated by the tendency of the current, and with the utmost ease enter and pass the opening of the draw. In a high stage of water, the boats do not regard the channel, as they find sufficient depth, generally, for navigation. Under such circumstances, they run very much from point to point, so as to make the shortest distance. At this locality, there is a general bend in the shore of the river, the convex of the curve being on the east side. In a downward passage, the tendency of the current is to force the boats toward the west side, especially at the point about three-quarters of a mile above the bridge, and the navigator is compelled to hold his rudder steadily against this influence, the result of which is an easy command of direction, to pass safely through the draw. He has no more bend to make than is necessary to conform to the shore of the river, and which he would aim to do, if there was no bridge to pass. Nor is there any difficulty or unusual operation in navigating a boat in this way. It must be kept in mind that there is no sudden turn in this case (nothing to compare with those that occur often in following the natural channel), but an easy bearing of the rudder in one steady direction against the slight tendency of current; and I am fully satisfied that any skillful navigator may run his boat to a particular object with more certainty than he would in the direct thread of the stream. In the latter case, the rudder would be required to alternate more or less from opposite directions, and the uniform, steady command would be less perfect than in the former.

The approach to the draw from below seems free from difficulty. Boats can easily take a position to enter it in perfect line; and when alongside the guide-pier, the liberal opening gives the power to pass up the river as the navigator may choose. You ask my opinion as to what can be done to improve the navigation at the draw? In answer, I do not see the necessity of anything. It appears to my judgment there is no difficulty in navigating it with perfect safety as it is. You can provide no security against unskillful or careless navigators, or insufficient machinery. I have seen many draws, where triple the navigation is made, far less conveniently arranged for the navigator, that have been years in operation, and no one thinks of complaining. It is only recently that so large openings have been required for draws. Formerly the opening was only wide enough just to allow the vessel to be warped through. Your draw allows a steamboat to pass under full headway, and an easy, spacious berth for that.

Under no circumstances, hardly, can a river be bridged and a draw provided, so as to leave the navigation as perfectly free as without a bridge. But the navigation may be substantially provided for, and every material interest secured by such a draw as you have erected. This question of drawbridges involves the accommodation, not only of navigation, but also the passage of railroad trains, as two distinct means of transportation. The latter has recently come into use. Formerly natural navigation, being the only means of cheap transport, was regarded as a thing too sacred to be put to the least inconvenience. Now, the railway has produced "an epoch in the affairs of mankind," and navigation is no longer the only means of easy intercommunication among men. Other methods equally demand the consideration of the public ; and hence the navigation must inevitably yield so much as will allow the enjoyment of the other. It has no natural claim to monopoly— not any claim, except so far as it affords public convenience; and this must necessarily be effected by other objects of public convenience. It is a favorable circumstance to the navigation that the drawbridge so amply provides for it. The railroad is also provided for, by this means, while it bears greater inconvenience than the navigation, in its use, and is moreover subject to all the expense of construction, maintenance and attendance. The railway is yet of moderate extension west of the river; but its progress will be onward, and every year will extend the area of country that will enlarge its traffic, until it will far exceed that of the navigation at this point. It is a mode of transport open at all seasons of the year, affording a vast superiority in the regularity and certainty of its operations. The navigation is closed a large portion of the year by frost, and greatly impeded; and, at times, arrested by low water during the remainder of the year. As the country draining into the Mississippi becomes settled and cultivated, this river must share the fate of all rivers, namely, the loss from increased absorption and evaporation, which surely reduces the depth of water in the low stages, and protracts the duration of these stages, tending materially to impair the value and importance of the navigation. The trade that heretofore passed on the Mississippi, is now being diverted by railroads, that reach it from the east, at Alton, Burlington, Rock Island, Fulton, Dubuque, and St. Louis. Other railroads are in progress to intercept and divert from other points more or less of the traffic that heretofore had no other outlet but the river. Some of these roads have proceeded west of the river, and probably the others soon will. The result of these railroads on the irregular and uncertain trade of the river cannot be a matter of doubt. Certain kinds of trade, and peculiar conditions of market, will doubtless perpetuate a considerable amount of river traffic. But the former special importance of the navigation has passed away, and it is now merely a competitor with the railroads, and entitled to no greater share of public consideration. It is no presumption to say, that in a few years either of the main lines of railroads above the Missouri will have a larger traffic than the Mississippi above the confluence of these rivers. It would, indeed, be a poor compliment to the discrimination of the public mind, to suppose they would not appreciate their interest in the great lines of railway that are now just beginning to bear the mighty volume of trade and intercourse between the vast prairies west of the Mississippi and the eastern market. There are interests connected with the navigation, and opposed to the diversion of this interest by railroads, or any other mode of communication. In reality, this navigation interest has nothing worth complaint in the matter of inconvenience caused by a drawbridge. The real matter of dissatisfaction must be sought in the inevitable diversion of trade produced by the railroads. The navigation must lose its relative importance, and, instead of enjoying a monopoly, must be content to be a competitor. It may be, however, that the great growth of the trade that will be caused by the railroads, in the rapid and extended settlement of the country, will still keep up, and perhaps increase, the business of navigation. The fraction it may obtain may be greater than its present whole. The public are interested that no unnecessary impediment shall obstruct either channel of transport, and especially the one that is likely to afford the most regular and reliable means of communication. The time is not distant when t will be regarded as a favor to the navigation, to provide for it ample and convenient draws to pass railroad bridges.

Respectfully submitted,

JOHN B. JERVIS,

Civil Engineer.

ADDENDUM.

The facts set forth in the annexed statement, entitled "RECORD OF STEAM-BOATS *passing through the draw of the Railroad Bridge across the Mississippi River at Rock Island,*" &c., go to show that the traffic of the Railroad, that *now* passes the Bridge at Rock Island, is more important than the trade of the River at the same place. If this result has been reached by the trade of sixty-six (66) miles of Railroad, in its barely introductory traffic, the extension of the roads across the State of Iowa must make the comparative importance greatly beyond that of the river. They not merely sustain the conclusions in the foregoing report, as to the relative importance of the two modes of inter-communication, but show that the railroad trade will greatly transcend, in importance, that of the river ; and, consequently, that the public are deeply interested in the maintenance of the bridge, as indispensable to the proper and convenient accommodation of the large traffic which the railroads are sure to accumulate at this place.

As the facts above referred to have come to my knowledge since my report was submitted, it has appeared proper to notice them, not merely as corroborating the conclusions of my report, but as eminently calculated to illustrate the influence and importance of railroads on the wealth and civilization of those vast and fertile prairies of the West, now secluded and destitute of any regular and cheap means of communication. It is no longer necessary, in this case, to reason wholly from analogy.

This line of Railroad, though in its infancy, and scarcely completed in the short sections that have been put in operation, presents, by a few months' operation, a state of facts which indicate, beyond question a traffic, that will leave far behind the *River trade* at Rock Island.

JOHN B. JERVIS,
C. E.

New York, January 20th, 1857.

Record of Steamboats passing through the Draw of the Railroad Bridge across the Mississippi River, at Rock Island, from the time it was repaired, August 4, 1856, to the close of Navigation, November 29, 1856.

	1	2	3	4	5	6	7	8	9	10	11	12	13	14	15	16	17	18	19	20	21	22	23	24	25	26	27	28	29	30	31
August.																															
Boats bound up	1				1	2	1	1	1		1	2				2	1				1	1	1	1	1	2			1	1	1
" " with 1 barge		2	1							2		2				1	1			1		1		1	1		1	2		1	1
" " 2 "	1	1																													
" down	1			1	1			2	1	3	3								1												1
" " with 1 barge								1	1	1	1																				
" " 2 "																															
September.																															
Boats bound up			1			1				2		2	2	1				2			1		2	1	1	1	1			1	
" " and 1 barge			1			1			1									1					2	1	1	1			1	1	
" " 2 "																															
" down				3					2		3													1							
" " and 1 barge				2					1																						
" " 2 "				1																											
October.																															
Boats bound up		1	1	1	2	1			1	1			1		2	1				2		1		1		1	1		1	1	
" " and 1 barge			1		1	1			1	1						1				1		1		2		1				1	
" " 2 "																						2									
" down	1			1	2				2				1					1										2			
" " and 1 barge			1						1																						
" " 2 "																															
November.																															
Boats bound up			1		2	1		2	4		2			1			1	1	1		1				2	1					
" " and 1 barge	1				1				2		1		1	1					3			2			1	1			2		
" " 2 "					1						3			6															1		
" down	1	1			1			1						1			2	1			1	1									
" " and 1 barge								2					2																		
" " 2 "								2																							

RECAPITULATION.

	Of this number there were 39 boats up without barges.	
84 boats bound up.	27 " " with 1 barge.	61 boats down without barges.
78 " " down.	18 " " 2 "	10 " " with 1 barge.
		7 " " 2 "

162 Total number of passages.

Of this number not one boat came in contact with any portion of the bridge or pier, and but one barge connected with a boat, and that not to do any injury, to my knowledge.

SETH GURNEY.

STATE OF ILLINOIS, } ss
Rock Island County. }

Seth Gurney, being duly sworn, states, that the matters and things contained in the foregoing statement are true to the best of his knowledge and belief.

[SEAL.] In testimony whereof, I have hereunto set my hand and notarial seal, this 1st day of January, A.D. 1857.

R. WILKINSON,
Notary Public.

Statement of the number of Cars crossing the Railroad Bridge from September 8th, to December 31st, 1856, both days inclusive, being 115 days, viz.:

367 baggage cars.	Total amount of Freight crossed, 33,886 Tons.
771 passenger cars.	Total number of Passengers crossed, 26,788
4,841 loaded freight cars.	

Total number of cars, 5,979, exclusive of empty Freight Cars returned.

[Intentionally Left Blank]

[Intentionally Left Blank]

[Intentionally Left Blank]

[Intentionally Left Blank]

[Intentionally Left Blank]

[Intentionally Left Blank]

Printed in Dunstable, United Kingdom